THE COMPREHENSIVE GUIDE TO PROFITABLE WORM FARMING

Cultivating Wealth from the Ground Up

ROBERTO M. GASPAR

Copyright © 2024 by Roberto M. Gaspar

TABLE OF CONTENTS

TABLE OF CONTENTS —————————————— 2-4

INTRODUCTION ————————————————— 5-7

CHAPTER 1 ————————————————————— 8-14

Overview of Worm Farming

Importance of Sustainable Agriculture

Getting Started: Setting Up Your Worm Farm

Setting Up Your Worm Farm

Choosing the Right Worm Species

CHAPTER 2 ———————————————————— 15-23

Understanding Worms: Worm Biology and Behavior

Worm Biology

Worm Behavior

Creating the Ideal Habitat

Feeding and Nutrition: Proper Diet for Worms

Proper Diet for Worms

Nutrient-Rich Feed Options

CHAPTER 3 ———————————————————— 24-32

Managing the Farm: Temperature and Moisture Control

Temperature Control

Moisture Control

Troubleshooting Common Issues

Harvesting and Processing: Harvesting Techniques

Harvesting Techniques

Processing Worm Castings for Maximum Benefit

CHAPTER 4 ----------------------------------- 33-42

Marketing and Sales: Identifying Target Markets

Identifying Target Markets

Effective Sales and Marketing Strategies:

Financial Management: Budgeting for a Worm Farm

Budgeting for a Worm Farm

Profitability Analysis

CHAPTER 5 ----------------------------------- 43-52

Scaling Up: Expanding Your Worm Farming Operation

Expanding Your Worm Farming Operation

Implementing Sustainable Practices

Success Stories in Worm Farming

Lessons Learned from Experienced Worm Farmers

CHAPTER 6 ————————————————————53-58

Frequently Asked Questions: Common Queries and Answers

CONCLUSION ——————————————— 59-61

APPRECIATION ————————————— 62-64

INTRODUCTION

Welcome to the fascinating world of "The Comprehensive Guide to Profitable Worm Farming" – your passport to unlocking the boundless potential beneath the surface. Have you ever envisioned turning a simple, sustainable practice into a flourishing venture? This guide is your compass, navigating you through the intricacies of worm farming and opening doors to financial success.

Embark on a Journey of Discovery: In these pages, we delve into the extraordinary realm of worm farming, transforming what might seem like a humble pursuit into a strategic and lucrative business opportunity. From the essential foundations of setting up your worm farm to the intricacies of harvesting and processing, each chapter is a stepping stone toward realizing the full potential of your agribusiness venture.

Why Worm Farming?

Ask yourself: Can the soil beneath your feet be the foundation for your financial growth? The answer lies in the transformative power of worm farming. Discover the eco-friendly advantages, the sustainable practices, and the endless possibilities for profit that this overlooked industry holds.

What Awaits You:

- Uncover the secrets of worm biology and behavior.

- Learn the art of creating an optimal habitat for thriving worms.

- Navigate the intricacies of nutrient-rich feed options for optimal growth.

- Master the techniques of harvesting and processing worm castings.

- Explore proven marketing and sales strategies to reach your target audience.

- Gain financial wisdom for a solid foundation and sustainable growth.

As you embark on this journey, consider this guide not just a manual but a companion – walking you through the exciting and rewarding path of turning worms into wealth. Are you ready to cultivate success from the ground up? Your journey begins now. Welcome to "The Comprehensive Guide to Profitable Worm Farming."

CHAPTER 1

Overview of Worm Farming

Worm farming, scientifically known as vermiculture, is a sustainable agricultural practice that harnesses the transformative power of earthworms. These humble creatures play a pivotal role in enhancing soil health and fertility. In a controlled environment, such as a worm farm, they efficiently break down organic matter, producing nutrient-rich castings that serve as a potent natural fertilizer.

The process of worm farming involves creating a suitable habitat for the worms, understanding their biological needs, and providing them with an organic diet. As they digest and decompose organic material, they release valuable nutrients, enzymes, and beneficial microorganisms into the soil. Worm farming is not only a practical solution for waste management but also a method that promotes healthy plant growth and overall ecosystem balance.

Importance of Sustainable Agriculture

Sustainable agriculture is a holistic approach that seeks to address the current and future needs of our planet while minimizing environmental impact. In the context of worm farming, sustainability takes center stage. Unlike conventional farming practices that may deplete soil nutrients and rely heavily on synthetic inputs, sustainable agriculture, with the incorporation of practices like vermiculture, focuses on maintaining soil fertility, conserving water, and reducing chemical use.

Worm farming contributes to sustainable agriculture by enhancing soil structure, promoting water retention, and fostering biodiversity. The nutrient-rich castings produced by worms improve soil fertility naturally, reducing the dependence on synthetic fertilizers. This not only benefits crop yields but also mitigates the environmental footprint associated with traditional farming methods.

In essence, the marriage of worm farming and sustainable agriculture is a harmonious alliance that not only nurtures the health of the soil but also aligns with our collective responsibility to protect the environment. By understanding the intricacies of worm farming and embracing sustainable practices, we pave the way for a more resilient and ecologically balanced agricultural future.

Getting Started: Setting Up Your Worm Farm

Embarking on the journey of worm farming is an exciting venture that promises not only a sustainable hobby but a potential source of income. Getting started involves laying the foundational steps for a successful and thriving worm farm. Here's a guide to setting up your worm farm and ensuring it becomes a thriving ecosystem.

Setting Up Your Worm Farm

Container Selection: Begin by choosing a suitable container for your worm farm. Opt for containers with good drainage, ventilation, and insulation. This could range from specially designed worm bins to repurposed containers such as wooden boxes or plastic bins.

Bedding Material: Create a comfortable habitat for your worms by providing the right bedding material. Common materials include shredded newspaper, cardboard, coconut coir, or a combination of these. Ensure the bedding is moist but not waterlogged.

Adding Food Scraps: Introduce kitchen scraps such as fruit and vegetable peels, coffee grounds, and eggshells to provide a nutrient-rich diet for your worms. Avoid adding meat, dairy, or oily items, as these can attract pests and create an unpleasant environment.

Introducing the Worms: Once the bedding and food scraps are in place, it's time to introduce the stars of the show – the worms. Choose a reputable source for acquiring your worms, and gently introduce them into the bedding. Red wiggler worms (Eisenia fetida) are a popular choice for their efficiency in breaking down organic matter.

Monitoring and Adjusting: Regularly monitor the moisture levels, temperature, and overall health of your worm farm. Adjust the conditions as needed to ensure a comfortable and productive environment for your worms.

Choosing the Right Worm Species

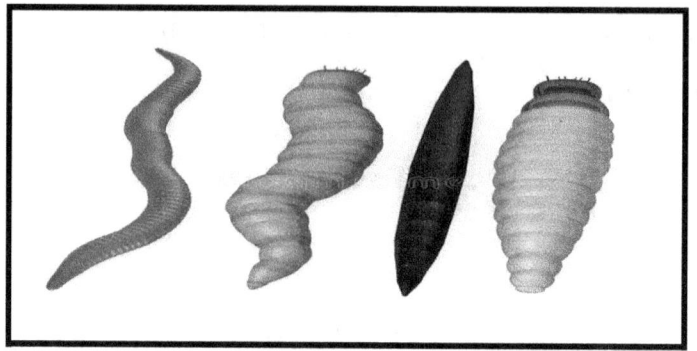

1. Red Wigglers (Eisenia fetida): These worms are well-suited for composting and thrive in the confined space of a worm bin. They have voracious appetites and efficiently break down organic matter.

2. European Nightcrawlers (Eisenia hortensis): Known for their larger size compared to red wigglers, European nightcrawlers are excellent for aerating the soil. They are more tolerant of cooler temperatures.

3. African Nightcrawlers (Eudrilus eugeniae): Ideal for warmer climates, African nightcrawlers excel in breaking down organic material quickly. They are known for their robust appetite and adaptability.

Choosing the right worm species depends on factors such as your location, the intended purpose of your worm farm, and the specific conditions you can provide. By carefully setting up your worm farm and selecting the appropriate worm species, you lay the groundwork for a successful and rewarding journey into vermiculture.

CHAPTER 2

Understanding Worms: Worm Biology and Behavior

 To truly master the art of worm farming, delving into the intricate world of worm biology and behavior is paramount. A comprehensive grasp of these aspects enables you to create conditions that optimize their well-being and productivity.

Worm Biology

1. Anatomy: Earthworms possess a segmented body with a head, tail, and a multitude of tiny bristles called setae. Understanding their anatomy helps in appreciating their role in the ecosystem.

2. Reproduction: Earthworms are hermaphrodites, possessing both male and female reproductive organs. They engage in a fascinating mating ritual, exchanging sperm with other worms during a copulatory dance.

3. Digestive System: Worms lack teeth but employ strong muscular contractions to break down organic matter. Their digestive system produces nutrient-rich castings, a valuable byproduct sought after for its fertilizing properties.

Worm Behavior

1. Photophobia: Worms are sensitive to light and tend to retreat from it. This behavior is crucial to understand when managing and harvesting your worm farm.

2. Thigmotaxis: Worms exhibit thigmotactic behavior, meaning they seek contact with surfaces. This behavior influences their movement and is important to consider when designing their habitat.

3. Escape Responses: Worms have a natural instinct to escape unfavorable conditions. Maintaining suitable conditions within the worm farm prevents them from attempting to flee.

Creating the Ideal Habitat

1. Moisture Control: Worms require a moist environment for respiration through their skin. Ensure the bedding remains consistently damp but not waterlogged to facilitate their well-being.

2. Temperature Considerations: Worms thrive in a temperature range of 55-77°F (13-25°C). Extreme heat or cold can stress the worms, impacting their activity and reproduction.

3. Proper Bedding: Choose suitable bedding materials such as shredded newspaper, coconut coir, or cardboard. This provides both a physical environment and a carbon source for the worms.

4. Adequate Ventilation: Worms need oxygen, and proper ventilation in the worm bin ensures they have a continuous supply. Well-designed bins include ventilation holes or a breathable design.

Understanding the intricacies of worm biology and behavior is the key to creating an environment where they can flourish. By tailoring the habitat to meet their needs, you not only ensure the well-being of your worms but also unlock their full potential in converting organic waste into nutrient-rich castings. This knowledge forms the foundation for a successful and sustainable worm farming venture.

Feeding and Nutrition: Proper Diet for Worms

Ensuring a balanced and nutritious diet for your worms is pivotal in maintaining a healthy and productive worm farm. By understanding the ideal foods and creating a well-rounded feeding regimen, you optimize their digestion process and enhance the quality of the resulting vermicompost.

Proper Diet for Worms

1. Fruit and Vegetable Scraps: Provide a variety of fruit and vegetable scraps such as apple cores, banana peels, carrot tops, and cucumber ends. These kitchen scraps are not only abundant but also offer essential nutrients for worm health.

2. Coffee Grounds: Rich in nitrogen, coffee grounds are an excellent addition to the worm diet. However, use them in moderation to avoid creating an overly acidic environment in the worm bin.

3. Eggshells: Crushed eggshells contribute calcium to the worm farm, promoting healthy reproduction and overall well-being. Rinse them thoroughly before adding to the bin.

4. Shredded Paper or Cardboard: Serve as a carbon source in the worm diet, helping to balance the carbon-to-nitrogen ratio. Shredded newspaper or cardboard also maintain proper bedding consistency.

5. Avoid Meat and Dairy Products: Refrain from adding meat, dairy, oily, or excessively acidic foods to the worm bin. These items can attract pests, disrupt the balance of the bin, and lead to unpleasant odors.

Nutrient-Rich Feed Options

1. Manure: Incorporate well-aged manure into the worm diet for added nitrogen and a diverse range of beneficial microorganisms. Ensure the manure comes from herbivores and has undergone proper composting.

2. Fall Leaves: Rich in carbon, dried and shredded fall leaves provide an excellent balance to the nutrient content in the bin. They also add structure to the bedding.

3. Vegetable Peels: Nutrient-dense vegetable peels, such as those from potatoes and carrots, offer a source of vitamins and minerals. Chop them into smaller pieces for easier consumption.

4. Herbivore Animal Bedding: If you have access to bedding material from herbivore animals like rabbits or guinea pigs, it can be added to the worm bin as both bedding and a supplementary food source.

5. Compostable Plant Trimmings: Non-diseased plant trimmings from your garden or kitchen garden provide a diverse array of nutrients and contribute to a well-balanced diet.

Maintaining a varied and well-managed diet for your worms is the key to their optimal health and productivity. By balancing the carbon and nitrogen sources and avoiding potential harmful items, you set the stage for a thriving worm farm that consistently produces nutrient-rich castings.

CHAPTER 3

Managing the Farm: Temperature and Moisture Control

 Effective management of your worm farm involves meticulous control of both temperature and moisture levels. These factors play a crucial role in creating an environment where worms thrive, reproduce, and efficiently convert organic matter into nutrient-rich castings.

Temperature Control

1. Optimal Range: Worms flourish in temperatures ranging from 55-77°F (13-25°C). Maintain this temperature range to ensure their metabolic processes operate optimally.

2. Avoid Extremes: Extreme heat or cold can stress worms and impede their activity. Protect the worm farm from direct sunlight, and if located in a colder climate, consider insulating the bin during winter.

3. Monitoring: Regularly monitor the temperature within the worm farm. Use a thermometer to ensure it stays within the preferred range. Consider relocating the bin during extreme weather conditions.

Moisture Control

1. **Consistent Moisture:** Worms require a consistently moist environment for respiration through their skin. Regularly check the moisture levels in the bedding. It should feel like a wrung-out sponge.

2. **Avoid Waterlogging:** While moisture is crucial, avoid waterlogging, which can suffocate the worms. Adjust the bedding and drainage to maintain an optimal balance.

3. **Water Source:** If the bedding becomes too dry, provide a water source by misting or adding moist food scraps. Conversely, if it's too wet, add dry bedding material to absorb excess moisture.

4. **Covering the Bin:** A well-fitted lid helps regulate moisture levels by preventing excess evaporation. It

also shields the worms from extreme weather conditions.

Troubleshooting Common Issues

1. Unpleasant Odors: Foul smells indicate imbalances in the bin. Adjust the carbon-to-nitrogen ratio by adding more bedding, and avoid overfeeding to prevent anaerobic conditions.

2. Pests: Fruit flies or other pests can be attracted to the worm farm. Bury food scraps deeper into the bedding, cover with additional bedding, and ensure proper ventilation to deter pests.

3. Worm Migration: If worms are attempting to escape, assess the conditions within the bin. Check for excessive moisture, acidity, or other stressors. Make adjustments accordingly.

4. Slow Decomposition: If the decomposition process seems slow, evaluate the types of food provided. Ensure a diverse diet and proper bedding consistency to stimulate microbial activity.

Effective farm management involves being attuned to the needs of your worm community. By maintaining optimal temperature and moisture levels and promptly addressing common issues, you create an environment conducive to the well-being and productivity of your worms. Regular observation and thoughtful adjustments ensure a thriving and sustainable worm farming venture.

Harvesting and Processing: Harvesting Techniques

As you witness the flourishing activity within your worm farm, the anticipation of harvesting nutrient-rich castings becomes a pivotal step in reaping the rewards of your efforts. Employing effective harvesting techniques ensures a smooth extraction process while maintaining the health of your worm community.

Harvesting Techniques

1. Selective Harvesting: Gradually move fresh food and bedding to one side of the bin, prompting worms to migrate to the new material. This allows you to scoop out well-populated sections of castings while leaving the majority of worms behind.

2. Light Exposure: Take advantage of worms' aversion to light. Expose the top layer of the bin to light, and as the worms burrow deeper to escape, scoop out the castings from the surface.

3. Vertical Migration: Utilize a vertical migration system by placing fresh bedding and food at the top of the bin. Worms naturally migrate upwards, making it easier to collect castings from the lower layers.

4. Use of Screens: Employ screens or mesh to separate worms from the castings. Place the material on the screen and allow worms to move through, leaving the castings behind.

5. Rest Periods: Implement rest periods where you stop adding new food for a week or two. This encourages the worms to finish processing existing material, making it simpler to harvest castings.

Processing Worm Castings for Maximum Benefit

1. Screening and Sifting: After harvesting, screen the castings to remove any remaining debris or unprocessed material. This fine sifting ensures a high-quality, homogeneous product.

2. Curing Period: Allow the harvested castings to undergo a curing period. This additional time enhances the nutrient concentration and microbial activity, resulting in a more potent organic fertilizer.

3. Storage in Ventilated Containers: Store processed worm castings in well-ventilated containers to maintain their quality. Avoid airtight containers to prevent moisture buildup, which can lead to clumping.

4. Dilution for Liquid Form: For liquid applications, dilute worm castings with water to create a nutrient-rich liquid fertilizer. This allows for easier application and absorption by plants.

5. Incorporate into Potting Mix: Blend worm castings into potting mixes to enrich the soil with essential nutrients. This enhances the growth and vitality of your plants.

Harvesting and processing worm castings are pivotal steps in converting the efforts of your worm farming venture into a valuable resource. By employing these techniques, you not only extract a high-quality organic fertilizer but also ensure the continued health and productivity of your worm community. This transformative process turns waste into wealth, fostering a sustainable and nutrient-rich environment for your plants.

CHAPTER 4

Marketing and Sales: Identifying Target Markets

 Successfully bringing your worm farming venture to the market hinges on a thorough understanding of your target audience. Identifying specific markets tailored to your products ensures a more focused and impactful marketing approach, paving the way for sustained growth and profitability.

1. Gardening Enthusiasts: Position your worm castings and related products towards gardening enthusiasts, including home gardeners, urban gardeners, and those passionate about sustainable gardening practices.

2. Organic Farms and Agriculture: Target organic farms and agriculture businesses looking for natural and eco-friendly fertilizers. Emphasize the organic and sustainable aspects of your worm castings.

3. Landscaping Companies: Collaborate with landscaping companies that prioritize soil health and sustainable landscaping practices. Your products can enhance the vitality of lawns, gardens, and outdoor spaces.

4. Nurseries and Garden Centers: Establish partnerships with nurseries and garden centers where customers frequently seek high-quality soil amendments. Provide them with information on the benefits and applications of your worm castings.

5. Educational Institutions: Market your products to schools, colleges, and educational institutions with agricultural programs. Position worm farming as an educational tool for sustainable agriculture.

Effective Sales and Marketing Strategies:

1. Online Presence: Develop a professional and user-friendly website showcasing your products, benefits, and ordering information. Leverage social media platforms to engage with your audience and share educational content.

2. Educational Content: Create informative content, such as blog posts, videos, or webinars, highlighting the benefits of worm castings and providing gardening tips. Position yourself as an authority in the field.

3. Product Packaging and Labeling: Invest in attractive and informative packaging that clearly communicates the benefits of your worm castings. Include instructions for use and certifications related to organic and sustainable practices.

4. Customer Testimonials: Encourage satisfied customers to provide testimonials and reviews. Positive feedback builds trust and credibility, influencing potential buyers.

5. Networking and Collaborations: Attend gardening and agriculture events, trade shows, and networking gatherings. Collaborate with influencers, gardening experts, and sustainable agriculture advocates to expand your reach.

6. Seasonal Promotions: Offer seasonal promotions and discounts to capitalize on peak gardening periods. Create bundled packages or limited-time offers to incentivize purchases.

7. Local Community Engagement: Engage with your local community through farmers' markets, community events, or workshops. Establishing a presence locally builds a loyal customer base.

8. Certifications: Obtain relevant certifications for organic and sustainable practices. Display these certifications prominently in your marketing materials to appeal to environmentally conscious consumers.

By identifying specific target markets and implementing effective sales and marketing strategies, you position your worm farming venture for success. Engage with your audience, provide valuable information, and emphasize the unique benefits of your products to create a lasting impact in the agricultural and gardening communities.

Financial Management: Budgeting for a Worm Farm

Prudent financial management is the bedrock of a successful worm farming venture. By crafting a well-thought-out budget, you lay the foundation for sustainable operations, ensuring that resources are allocated efficiently and that your business remains financially resilient.

Budgeting for a Worm Farm

1. Infrastructure Costs: Include expenses related to setting up the worm farm, such as containers, bedding materials, and any necessary infrastructure modifications. Consider both initial setup costs and ongoing maintenance expenses.

2. Worm Stock and Feed: Budget for acquiring your initial worm stock and ongoing costs for feed materials. Factor in fluctuations in pricing and availability, and ensure a reliable supply chain for consistent farming operations.

3. Operational Costs: Account for day-to-day operational expenses, including utilities, labor, and any additional materials required for maintaining the farm. Regularly review and adjust these costs to maintain efficiency.

4. Marketing and Promotion: Allocate a portion of your budget for marketing and promotional activities. This includes online and offline campaigns, packaging materials, and any collaborations with influencers or partners.

5. Education and Training: If applicable, budget for ongoing education and training. Staying informed about the latest developments in vermiculture and sustainable agriculture is essential for long-term success.

6. Contingency Fund: Establish a contingency fund to address unexpected expenses or fluctuations in market conditions. This financial buffer provides flexibility and resilience in the face of unforeseen challenges.

Profitability Analysis

1. Revenue Streams: Identify and analyze your primary revenue streams, such as sales of worm castings, worm stock, or related products. Diversify revenue streams where possible to mitigate risks.

2. Pricing Strategy: Evaluate your pricing strategy by considering production costs, market demand, and competitor pricing. Striking a balance between competitive pricing and maintaining profitability is crucial.

3. Cost of Goods Sold (COGS): Calculate your cost of goods sold, encompassing expenses directly tied to producing and delivering your products. This includes worm feed, labor, packaging, and distribution costs.

4. Gross Profit Margin: Determine your gross profit margin by subtracting COGS from total revenue and expressing the result as a percentage. This metric provides insights into the efficiency of your production process.

5. Operating Expenses: Analyze your operating expenses, including marketing, administrative costs, and utilities. Monitor and optimize these expenses to ensure they align with revenue growth.

6. Net Profit Margin: Calculate the net profit margin by subtracting all expenses, including operating expenses and taxes, from total revenue. This metric provides a comprehensive view of your overall profitability.

7. Return on Investment (ROI): Assess the return on investment for your worm farming venture. Evaluate the effectiveness of your budget allocations and make informed decisions for future investments.

By carefully budgeting for the various aspects of your worm farm and conducting a thorough profitability analysis, you gain insights into the financial health of your venture. This proactive financial management approach positions your worm farm for sustained growth and profitability in the dynamic agricultural market.

CHAPTER 5

Scaling Up: Expanding Your Worm Farming Operation

 As your passion for worm farming grows and the demand for your products increases, the prospect of scaling up your operation becomes an exciting endeavor. Scaling up involves strategic planning, infrastructure development, and the implementation of sustainable practices to ensure the continued success and resilience of your expanding worm farming venture.

1. Assess Market Demand: Before expanding, conduct a thorough assessment of market demand. Identify trends, analyze customer feedback, and evaluate potential new markets to ensure there's a sustainable demand for your products.

2. Infrastructure Planning: Plan for additional infrastructure to accommodate the expanded operation. This may include larger worm bins, additional storage facilities, and enhanced processing capabilities. Consider scalability in the design to accommodate future growth.

3. Worm Stock and Feed Supply: Ensure a reliable supply chain for acquiring additional worm stock and feed materials. Collaborate with suppliers to secure consistent access to quality resources as your operation expands.

4. Human Resources: Evaluate your staffing needs and consider hiring additional personnel to manage the increased workload. Provide training to ensure that new team members align with your farm's practices and values.

5. Technology Integration: Explore technology solutions to streamline processes and enhance efficiency. Implementing digital tools for monitoring, data analysis, and inventory management can significantly contribute to the smooth operation of an expanded farm.

Implementing Sustainable Practices

1. Eco-Friendly Infrastructure: Construct or upgrade infrastructure with a focus on sustainability. Utilize recycled materials, incorporate energy-efficient systems, and design facilities that minimize environmental impact.

2. Waste Reduction Strategies: Implement waste reduction strategies within your expanded farm. Optimize feeding practices to reduce excess organic matter, explore options for repurposing byproducts, and minimize overall waste generation.

3. Water Conservation: As you scale up, prioritize water conservation efforts. Implement rainwater harvesting systems, explore efficient irrigation methods, and ensure that water usage aligns with sustainable practices.

4. Renewable Energy Sources: Consider integrating renewable energy sources, such as solar or wind power, into your expanded operation. This not only reduces environmental impact but also contributes to long-term cost savings.

5. Biodiversity Enhancement: Foster biodiversity within your expanded farm by incorporating native plants, creating natural habitats, and adopting agroecological practices. A biodiverse environment enhances the overall health and resilience of your farm.

6. Community Engagement: Engage with the local community and stakeholders to foster a sense of sustainability. Share your commitment to eco-friendly practices, collaborate with like-minded businesses, and contribute to local environmental initiatives.

Scaling up your worm farming operation is not just about quantity; it's an opportunity to enhance the quality and sustainability of your practices. By carefully planning for expansion and incorporating eco-friendly measures, you not only meet the increasing demand for your products but also contribute to a more sustainable and resilient agricultural ecosystem.

Exploring success stories in worm farming provides invaluable insights into the potential and possibilities within this sustainable agricultural practice. These case studies highlight the achievements, challenges, and innovative approaches of successful worm farmers, serving as beacons of inspiration for those navigating the world of vermiculture.

1. Urban Farming Triumph: A case study from an urban worm farm showcases how a small-scale operation in a city setting can thrive. By creatively utilizing limited space, implementing vertical farming techniques, and establishing partnerships with local businesses for waste materials, this farm has become a model of urban sustainability.

2. Community-Driven Enterprise: Another success story revolves around a community-driven worm farming enterprise. This case study emphasizes the power of community engagement, where local residents actively participate in vermiculture workshops, contribute kitchen scraps, and purchase the resulting worm castings. The farm has not only created a sustainable business but also fostered a sense of community and environmental stewardship.

3. Innovative Product Diversification: A forward-thinking worm farmer successfully diversified their product line beyond traditional worm castings. By experimenting with value-added products such as worm tea, organic fertilizers, and specialty compost blends, they have captured a broader market share and established themselves as industry innovators.

4. Technology-Enhanced Farming: In a case study focused on technology integration, a worm farm embraced modern tools for monitoring and managing their operation. Automated systems for temperature control, moisture sensing, and inventory management have significantly improved efficiency and allowed for seamless scaling up.

Lessons Learned from Experienced Worm Farmers

1. Adaptability is Key: Experienced worm farmers emphasize the importance of adaptability. Markets, weather conditions, and customer preferences can change, and successful farmers are those who can adjust their strategies and practices accordingly.

2. Continuous Learning: The journey of worm farming is one of continuous learning. Successful farmers invest time in staying informed about the latest research, innovative practices, and emerging trends in vermiculture. This commitment to ongoing education contributes to their sustained success.

3. Sustainable Practices Pay Off: Lessons from experienced worm farmers underscore the long-term benefits of sustainable practices. From efficient resource management to biodiversity enhancement, prioritizing sustainability not only benefits the environment but also contributes to the overall resilience and profitability of the farm.

4. Customer Engagement Matters: Building strong relationships with customers is a common theme among successful worm farmers. Whether through educational content, community events, or personalized interactions, engaging with customers fosters loyalty and word-of-mouth referrals.

5. Start Small, Scale Wisely: Many successful worm farmers advocate for starting small and scaling up gradually. This approach allows for a more manageable learning curve, helps in building a solid foundation, and minimizes the risks associated with rapid expansion.

Case studies and lessons learned from experienced worm farmers serve as valuable resources for those embarking on their own vermiculture journey. By drawing inspiration from successful practices and learning from challenges overcome, aspiring worm farmers can navigate the complexities of this sustainable agricultural venture with greater confidence and success.

CHAPTER 6

Frequently Asked Questions: Common Queries and Answers

Embarking on the journey of worm farming often sparks a multitude of questions. Understanding the intricacies of vermiculture is crucial for success. Here are answers to some frequently asked questions to guide both beginners and experienced worm farmers on their quest for sustainable agribusiness.

1. What is vermiculture, and how does it differ from traditional composting?

Answer: Vermiculture is the practice of using earthworms to decompose organic waste, resulting in nutrient-rich castings. Unlike traditional composting, vermiculture relies on the digestive processes of worms to break down material, creating a more concentrated and valuable end product.

2. Which worm species are best for vermiculture?

Answer: Red wigglers (Eisenia fetida) are widely considered ideal for vermiculture due to their efficient feeding habits, adaptability to confined spaces, and rapid reproduction. European nightcrawlers (Eisenia hortensis) and African nightcrawlers (Eudrilus eugeniae) are also popular choices, each with unique characteristics suited to specific conditions.

3. What can be fed to worms, and what should be avoided?

Answer: Worms can be fed a variety of kitchen scraps, including fruit and vegetable peels, coffee grounds, and eggshells. Avoid feeding them meat, dairy, oily, or excessively acidic foods, as these can lead to imbalances and attract pests.

4. How do I prevent unpleasant odors in my worm farm?

Answer: Foul odors often indicate imbalances in the bin. To prevent unpleasant smells, avoid overfeeding, maintain a proper carbon-to-nitrogen ratio, and ensure adequate aeration. Regularly turning the bedding can also help aerate the contents and prevent anaerobic conditions.

5. Can I use the liquid collected from my worm bin as fertilizer?

Answer: Yes, the liquid collected, often referred to as worm tea, can be diluted and used as an organic fertilizer. This nutrient-rich liquid provides plants with essential minerals and beneficial microorganisms.

6. How do I harvest worm castings from my worm farm?

Answer: Harvesting worm castings involves techniques such as selective harvesting, light exposure, and vertical migration. By moving fresh bedding and food to one side of the bin, worms migrate, allowing you to scoop out well-populated sections of castings.

7. Is worm farming profitable, and how long does it take to see results?

Answer: Worm farming can be profitable, but results vary based on factors such as scale, market demand, and business strategies. Generally, it takes a few months to start seeing significant results, with production and sales increasing as the worm population grows.

8. How can I prevent pests in my worm farm?

Answer: Bury food scraps deeper into the bedding, cover with additional bedding, and ensure proper ventilation to deter pests. Avoid adding large quantities of oily or sugary foods, as these can attract unwanted pests.

9. Can I use newspapers as bedding for my worm farm?

Answer: Yes, shredded newspaper is a commonly used bedding material for worm farms. Ensure it's moist but not waterlogged, and mix it with other materials like cardboard, coconut coir, or leaves for a balanced and comfortable habitat.

10. Are there any specific laws or regulations for worm farming?

Answer: Regulations vary by location, so it's essential to research local laws and agricultural guidelines. In many places, worm farming for personal use is typically unrestricted, but commercial operations may require compliance with specific regulations related to waste management and agriculture.

These frequently asked questions and answers offer foundational knowledge for those starting or refining their journey into vermiculture. Continuous learning, adaptation, and hands-on experience will further enhance your expertise in the intricate world of worm farming.

CONCLUSION

In closing the pages of "The Comprehensive Guide to Profitable Worm Farming," you've not only gained knowledge but have embarked on a transformative journey into the world of sustainable agribusiness. As you reflect on the insights shared within these chapters, envision the possibilities that lie beneath the surface, rich, nutrient-dense possibilities that extend far beyond the boundaries of conventional farming.

Harvesting Success: Your newfound understanding of worm biology, farm management, and marketing strategies equips you with the tools to turn your passion for worm farming into a thriving enterprise. The road to financial success begins with the humble earthworm, and you now possess the wisdom to cultivate prosperity from the ground up.

Embrace Sustainable Prosperity: In the pursuit of wealth, remember the symbiotic relationship between responsible farming and a thriving planet. By embracing sustainable practices, you contribute not only to your own prosperity but also to the well-being of our environment. Your journey as a responsible entrepreneur has the power to inspire others and shape a brighter, greener future.

As you close this chapter, consider it not an end but a new beginning. Implement the strategies, embrace the challenges, and watch your worm farm flourish. Share your success with others, fostering a community of forward-thinking farmers dedicated to sustainable and profitable practices.

Thank you for joining us on this educational journey into the world of profitable worm farming. May your fields be fertile, your harvests bountiful, and your entrepreneurial spirit evergreen. Your success is not just a conclusion to this guide; it is the opening chapter of a prosperous and sustainable future.

Cultivate success, nurture sustainability, and thrive in the world of profitable worm farming!

APPRECIATION

Dear Readers,

As you conclude your exploration of "The Comprehensive Guide to Profitable Worm Farming," I extend my heartfelt gratitude for accompanying me on this journey into the fascinating realm of sustainable agribusiness. Crafting this guide has been a labor of passion, driven by the belief that beneath the surface lies not just soil but untapped opportunities for prosperity.

A Personal Insight: Worm farming, often underestimated, holds the potential to revolutionize agriculture. My own journey into this field has been one of continual learning, experimentation, and, most importantly, reaping the rewards of a sustainable and profitable endeavor. It is my sincere hope that the knowledge and insights shared within these pages ignite a similar passion within you.

Sustainability in Action: As stewards of the Earth, our responsibility extends beyond personal gain. By adopting sustainable farming practices, we contribute to the collective effort of preserving our planet. Worm farming, in its simplicity and effectiveness, exemplifies how small changes in our approach can yield substantial benefits for both our businesses and the environment.

Your Journey Continues: Remember, this guide is not just a manual but a companion on your journey to financial success. Embrace the challenges, implement the strategies, and adapt them to suit your unique circumstances. Your success story in worm farming adds another chapter to the narrative of sustainable entrepreneurship.

Connect with Me: I invite you to share your experiences, challenges, and triumphs. Let us build a community of passionate individuals dedicated to cultivating prosperity responsibly. To share your insights, seek advice, or simply celebrate your achievements.

Thank you for entrusting me with the opportunity to guide you in the world of profitable worm farming. May your farms thrive, your endeavors prosper, and your commitment to sustainability resonate with others.

Best Wishes,

Roberto M. Gaspar
Author, "The Comprehensive Guide to Profitable Worm Farming"